TIME

Continuity By Creation

Ricky Bennison

Published by Robin Publishing

TIME

Continuity By Creation

ISBN: 978-0-9933963-6-6

Front and back cover artwork: Two famous movers?

To Aunt Marie,

may you rest in peace.

'And she rode a white horse'

Table of Contents

Introduction

Continuity by creation is a logical theory which explains everyday observations which can be made in regard to time's continuity. A continuity which we all live by, and can identify and understand. Observable phenomenon such as the simultaneity of occurrence are explained by the unifying truth of creation. More specifically, by the self-creation of a new point of time which is the cause of time's progression. This explanation of time provides you with a framework with which to understand time's continuity, and also a solid grounding from which to explore the subject further.

Each point of the theory allows for much scope for contemplation. Some took me years to progress to. Without knowing the exact destination I was trying to reach in terms of understanding, my thoughts were extensive and by no means always correct. When I came to an accurate conclusion in regard to what an aspect of time was (as far as I could understand and linguistically articulate), I memorised it and used it as a waypoint to progress from. This was in addition to realising the value of the point in its own right. When I recalled these memorised statements they acted as memory triggers and brought to mind related points I had also thought of. This process of developing thought is partly why the theory uses the format of relatively concise statements that it does. On the one hand, it reflects the process of mental waypoint advancement which I used personally and this should help you to gradually progress your knowledge of how continuity by creation works in regard to time. On the other hand, understanding that you do not necessarily know what the related points are, it additionally gives you a solid foundation to expand upon. In basic terms, you can consider a stated point, expand upon it, consider it together with other points, and discover new points of understanding. These may be points of understanding which progress onto the originally considered points or new points which stem from them.

For example, we can consider the statement 'Everything which occurs in the present moment, occurs simultaneously.' and then apply it to an observation of the world around us. Upon observing present occurrence in the world we can see that everything is indeed occurring simultaneously. So the statement helps us to more fully consciously realise what we already instinctively, or subconsciously, know. It can thereby improve our understanding of time's nature. A more complex example can be found in the statement 'The only way there can always be something new, is by creation.' In this case we can consider why this might be so and involve other points in our consideration. Upon reading the theory in full, you will have seen how time is identified as being singular and its process self-relative. You will also have noticed it is identified as continuously progressing. As such, relative to itself, time, as in all time, would have already occurred as there would be no counteracting process to stop it fully occurring. On the basis it is continuous there would only be a reason for it to have fully occurred. There would not be anything to stop the full expression of its continuity, which without anything new being created would be an end. However, we know that time has not ended but continues. Any and all evidence demonstrates this. There must therefore always be something new to allow this continuity to occur: the new self-created point of time. It is the creation of this new point which causes time to be both a new and ongoing process. So by considering one of the stated points in conjunction with others, we have explored the nature of time further and enhanced our understanding. Certainly in terms of the depth of our conscious comprehension.

There are two layers to the theory in the book. A basic layer and a more advanced layer; the second layer building upon the first. The reason for this is that I thought it would be beneficial for you to gradually increase your understanding of the main points by revisiting them with the addition of connected ideas. Expanding on the theory in this way should help it to retain its clarity without too many strands being introduced too quickly.

It can be seen in the complementary theory of how the Universe increases in size whilst concurrently reordering itself (which may be considered as movement) that the logics of the two theories run an approximate parallel to each other. And this is not the only example. We can see in language how the addition of a new word, or sign, continues the previous communication while altering its meaning. To consider that the process of continuity by creation has readily observable physical manifestation in life is truly astounding. Perhaps the most beautiful example is the continuity of a family by the birth of a child. And the same is indeed true of any species of life: they all continue by creation. It is a fundamental theme of life and it is how time works.

Time (Basic)

This is a basic statement of points which relate to the nature of time, including two related focuses on the time continuum and the present moment.

The Time Continuum

1. The time continuum is singular. All time is part of the time continuum.

2. For something to be it must continue.

3. For something to continue it must continue to something, and for that something to be it must also continue to something, which must continue to something as well, and so forth in a continuous connection: a continuum.

    For something to be, it must also be present at itself; it must be present at the point where and when it is.

    For something to be, it must be both going to a point and be at a point simultaneously: to and at, at and to.

Considered in terms of the time continuum more generally, time is both at the new present moment and going to the new present moment simultaneously.

4. There is always a new point of time which all time continues directly in regard to.

5. The new point of time begins and the time continuum continues newly in regard to it.

6. The form something continues in always alters.

7. The time continuum continues because a new point of time is self-created. All time is unified with, continues in regard to and is defined by this point. This process is the present moment.

Aspects of the Present Moment of Time

1. The present moment is the process of the self-creation of a new point of time and all time continuing newly in regard to it, being unified with it and defined by it.

2. We are always in the present moment.

3. The present moment is always new.

4. The present moment is continuous.

5. That which occurs in the present moment, occurs instantaneously.

6. Everything which occurs in the present moment, occurs simultaneously.

7. Everything which is occurring, occurs at the same rate of time based occurrence.

8. Something's state of being is entirely defined by its present state.

9. The future is an idea which occurs in the present.

10. The past has occurred, and what constituted the past continues to occur in a newly defined form in the present. A memory occurs in the present.

Time (More advanced)

This is a more advanced statement of points which relate to the nature of time, including two related focuses on the time continuum and the present moment.

The Time Continuum

1. The time continuum is singular. All time is part of the time continuum.

2. For something to be it must continue.

3. For something to continue it must continue to something, and for that something to be it must also continue to something, which must continue too something as well, and so forth in a continuous connection: a continuum.

   Therefore, any given point of the continuum must continue to a point. To continue to, is part of its process of being, is part of what it does which is essential for it to be. In addition, any given point of the continuum must be at the place when it is; it must be present at itself; it is present at its own manifestation.

Therefore, to be at is also a part of its process of being. Considering the above together, for a thing to be, it must be both going to a point and be at a point concurrently: to and at, at and to. As such, any part of the continuum is both at itself and continuing to a new point simultaneously.

These processes are part of the same unified process of continuity. This logic is evidenced in life, by any given movement being both at a position and going to a new one simultaneously. It is important to note that everything in life always moves to a greater or lesser extent.

4. There is always a new point of time which all time continues directly in regard to. All points of the continuum continue solely in regard to this point. The only way there can always be something new, is by creation. The process of creation, and continuity in regard to it, constitutes the new moment of time, which is also known as the present moment.

5. The new point of time begins and the time continuum continues newly in regard to it. The continuity of time

in regard to the self-creation of the new point of time, occurs simultaneously with its creation.

Simultaneously with the creation of the new point of time, the previous moment of time ends; because what constituted what the previous moment of time was, now continues and is defined by the newest point of time. The previous ordering of things, by what was then the new point of time (the one which defined the previous moment), has ended, and therefore so has that moment of time.*

The new point of time is self-created before the previous moment of time ends, and while the previous moment ends in simultaneity with the occurrence of this process, its constituent elements continue in regard to the new self-created point of time.

The previous moment has ended because it is no longer defined by what was then the new point of time. What was previously the new point of time, is now redefined, along with all the constituent elements of the previous moment, by the newest self-created point of time, which defines the present moment.

There is only one defining new point of time, which defines the present moment; so while the constituent elements of the previous moment can continue as part of the present, the previous moment has ended.

The time continuum is a continuous process without a pause or stop. It does not start and then stop, or stop and then start; it is continuous without pause or gap. If a moment of the continuum was to end and then the next moment begin, this would mean that there was a period of cessation when the old moment ended. Time is a continuous progression of new moments, without stoppage, because the new moment begins before the old moment ends and therefore a continuity is maintained. The previous moment ends simultaneously with the self-creation of the new point of time, and the instantaneous creation of the present moment.

6. The form something continues in always alters. Because it is redefined by the new self-created point of time which it continues in regard to.

7.   The time continuum continues because a new point of time is self-created. All time is unified with, continues in regard to, and is defined by this new self-created point of time. This process is the present moment.

This was, in a previous version, 'the new point of time is self-created before the previous moment of time ends and therefore the previous moment does not end but continues in regard to the new self-created point of time.' However, it is not realistic to say the previous moment, and the past, continues as part of the present moment, even if its constituent elements do. This is, however, a relatively well pronounced example of how a mistake in philosophical theory can negatively affect a person's state of mind.

Aspects of the Present Moment of Time

1. The present moment is the process of the self-creation of a new point of time and all time continuing newly in regard to it, being unified with it and defined by it.

2. We are always in the present moment. The present moment of time is the one in which everything which exists is present in and defined by. All things which are occurring, occur in, and as part of, the present moment.

3. The present moment is always new.

4. The present moment is continuous. The present moment is always new and continuous. There is always a new point of time created and all time newly continues in regard to it. Everything continues in a manner which is determined by the newest point of time.

5. That which occurs, occurs instantaneously. If something is happening, it is happening instantaneously. For instance, a movement which is occurring in the present is occurring instantaneously.

6. Everything which is occurring, is occurring simultaneously. All that is occurring, occurs in unison with the new point of time, which instantaneously grants it continuity; therefore, all occurrence continues simultaneously.

7. Everything occurs at the same rate of time based occurrence. Whatever other values are ascribed to an occurrence, all that occurs, occurs at the same rate of time based continuity i.e. things continue to the next moment of time at the same time.

   This constant rate of progression, from one moment to the next which all things share, allows other values to exist in the same moment of time, and thereby interact with each other- in terms of their assessment, it allows them to be measured and compared. For instance, you can assess that a thing is moving faster than another thing because they are both moving at the same moment of time.

8. Something's state of being is entirely defined by its present state. What something is, is based upon what it is in the present moment; what it is now, as opposed to what it was or what it may be.

9. The future is an idea which occurs in the present.

   The only certainty of occurrence, is occurrence.

   The future is an idea, or other form of prediction, of something which may happen. The prediction may be considered to be more or less accurate, but will never exactly foretell what will happen. The future, as an idea, occurs as part of the present. For the future, as an actual moment of time, to be manifest at the same time as the present is not possible, because its manifestation would mean it was part of the present.

   However, due to the nature of the present's continuity- that it must always continue to something- the future can be considered to certainly occur; in that there must always be a future in order for there to be a present.

10. The past has occurred, and what constituted the past continues to occur in a newly defined form in the present. A memory occurs in the present.

Related Theories

The following section was written up several years prior to the 'Basic' and 'More advanced' sections found in this book. Whilst it is very similar in focus, I think that its approach and wording are quite different and so I have included it here instead of in the main body.

Time (Concise)

Time is the continuity of all things in regard to the self-creation of a new point of time. The new point of time is self-created ahead of the continuity of time. The self-creation of the new point of time creates the continuity of time. Time continues solely in regard to the self-creation of the new point of time, therefore time has one rate of continuity. The self-creation of the new point of time is eternal therefore time continues.

The following section is adapted from a related theory on how the Universe (which I also refer to as inistance) increases in size. The wording has been changed so time is referenced instead of the Universe. I thought that this would be an interesting way of exploring the relationship between the two theories- both of which are based upon the principal of creation.

The Time Continuum (considered in similar terms to the Universe)

The time continuum is singular. All time is a part of the time continuum.

The time continuum is the continuity of all time in regard to the self-creation of a new point of time.

The new point of time is self-created ahead of the continuity of time, which is simultaneously unified with it and therefore newly continues in regard to it. The new point of time is a part of the time continuum. The self-creation of the new point of time creates the continuity of time; time continues, and does not end, because it is unified with and continues in regard to the new self-created point of time.

Time continues directly, instantaneously and simultaneously, in regard to the new point of time and is newly defined by it. Time is entirely defined by the new self-created point of time.

The self-creation of the new point of time is eternal and therefore time continues.

This is the theory on how the Universe increases in size. Originally it was published in a book called *Alpha*. The term inistance (adapted from existence for various reasons related to semantics, breathing and pronunciation) is my preferred term; however, as the Universe is a more familiar term I also use it, and have done so here, for ease of comprehension. For similar reasons, extent is used here, where as intent (as in an inwardly orientated, or going, extent) is used in other versions in combination with inistance.

Christ All Mighty
He is born
Therefore we may live
King of Kings

The Universe

The Universe is singular. That which exists is part of the Universe.

The Universe is singular and has one direction. Towards its centre.

The Universe is singular because all points are positioned by the same centre. The inward direction of the Universe is created by being positioned by the centre.

The Universe has an inward extent which is the distance of its inward direction to and of its centre.

The inward extent of the Universe is increased by the self-creation of a new centre further inwards than the previous centre. The new centre is self-created ahead of the Universe. The self-creation of the new centre creates the inward extent of the Universe.

The self-creation of the new centre increases the inward extent of the Universe by an amount which is greater than the entire previous inward extent. The amount it is greater by is greater than the entire previous inward extent of the Universe.

All of the Universe is newly positioned by the new self-created centre.

The self-creation of the ever inward centre is eternal.

This was a section originally included in the book *Continuity by Creation* +. It helps to elucidate the process of how the Universe (initially referred to here as inistance) increases in size and how time continues. Importantly, it discusses both in regard to Christian theology.

Continuity by Creation: The Ever Inward Centre, The New Self-Created Point of Time, and Christian Theology

In a consideration of the nature of inistance and the ever inward centre, it is important to realise that whilst the ever inward centre is a constituent part of inistance, it is also the determining factor of inistance. Considering this in regard to the ever inward centres self-creation, inwardly ahead of inistance, it should be realised that the ever inward centre is internal because it orders inistance around itself (incorporating dimensional factors such as distance, direction and size) and not because it must be self-created inside inistance based upon a factor which exists prior to its self-creation.

Similarly, the self-creation of the new point of time is the cause and determining factor of the continuity of time.

The continuity of time does not cause the self-creation of the new point of time based upon a pre-determined process which means it must occur; the new point of time, which determines the continuity of time, freely chooses, based upon its own power, its own self-creation.

I think these two theories of continuity and increase, because of a self-created and all powerful principal, are similar to the Christian belief in the Immaculate Conception and the Resurrection of Jesus Christ, who is the saviour of humanity. To go further in this assessment, I think that the birth and Resurrection of Jesus Christ represents the salvation of all life, and I consider that this process is perhaps represented, albeit imperfectly, by the aforementioned theories on the continuity and increase of time and the Universe*; more specifically, that the ever inward centre, and the new self-created point of time, may be considered as aspects of the being of Jesus Christ.

On the basis of this interpretation, I do not think it is plausible that Jesus Christ died on the cross and was then resurrected from death; because if this was the case, and death was considered to be a cessation of life for whatever duration, then during that duration the Universe and time would end. I think that Jesus Christ was severely wounded whilst crucified on the cross, perhaps mortally, and may, or would, have died from his wounds, but was resurrected; He therefore did not die but was born anew. Due to the continuity of Christ's life, based upon His rebirth during His lifetime, life, the Universe, and time, could continue without stoppage.

For reasons of ease of comprehension, I will use the term 'the Universe' in preference to 'inistance' for this explanation.

Also available:

Continuity by Creation +

The Collected Writings of Ricky Bennison

Continuity by Creation + is a broad collection of ideas which includes theories on the nature of time, the Universe, mental focus, breathing, astronomy, and how energy navigates and moves internally.

The secret to happiness is freedom,
and the secret to freedom is courage.

~Thucydides